BUILDING BLOCKS OF MATH

UNITS OF MEASUREMENT

Written by Regina Osweiller

Illustrated by Daniel Hawkins

a Scott Fetzer company
Chicago

World Book, Inc.
180 North LaSalle Street
Suite 900
Chicago, Illinois 60601
USA

For information about other World Book publications, visit our website at **www.worldbook.com** or call **1-800-WORLDBK (967-5325)**.
For information about sales to schools and libraries, call 1-800-975-3250 (United States), or 1-800-837-5365 (Canada).

Library of Congress Cataloging-in-Publication Data for this volume has been applied for.

Building Blocks of Math
ISBN: 978-0-7166-4253-4 (set, hc.)

Units of Measurement
ISBN: 978-0-7166-4262-6 (hc.)

Also available as:
ISBN: 978-0-7166-4272-5 (e-book)

Printed in India by Thomson Press (India) Limited, Uttar Pradesh, India
2nd printing June 2023

Acknowledgments:
Writer: Regina Osweiller
Illustrator: Daniel Hawkins/The Bright Agency
Colorist: Leo Trinidad/The Bright Agency
Series Advisor: Marjorie Frank
Special thanks to KnowledgeWorks Global Ltd.

TABLE OF CONTENTS

THE IMPORTANCE OF MEASUREMENT
Dad, we're learning about measurement in school. Why is measurement important?
I'm glad you asked, son. I have a great measurement story that I want to read to you.
It's got wizards, a curious dog, a woman who travels in a soap bubble, and other fascinating characters ...
The WIZARD of MEASUREMENT
Dorothy Fourteen lives on a farm in Kansas. The family uses measurement units that are common in the United States.
Dorothy, please measure 8 pounds of scratch grains for the chickens.
Yes, Aunt M.
And we need water heated to 100 °F for the baby goats.
Yes, Uncle H.

Dorothy, can you help me cut wood posts of the length 4 feet, 6 inches? We need to fix the fence.
There are so many kinds of measurements! How do I keep track of them all?
Somewhere, above the clouds, I can learn about measuring. Length, weight, and temperature, too. It will be so reassuring.
Dad, where did you find this story? And do you really have to sing?
Just pay attention.
A mobile, destructive vortex of winds rotating at speeds up to 250 miles per hour is approaching from a mere 800 feet away!
What?!?!
It's a tornado!
WHOOOOSH

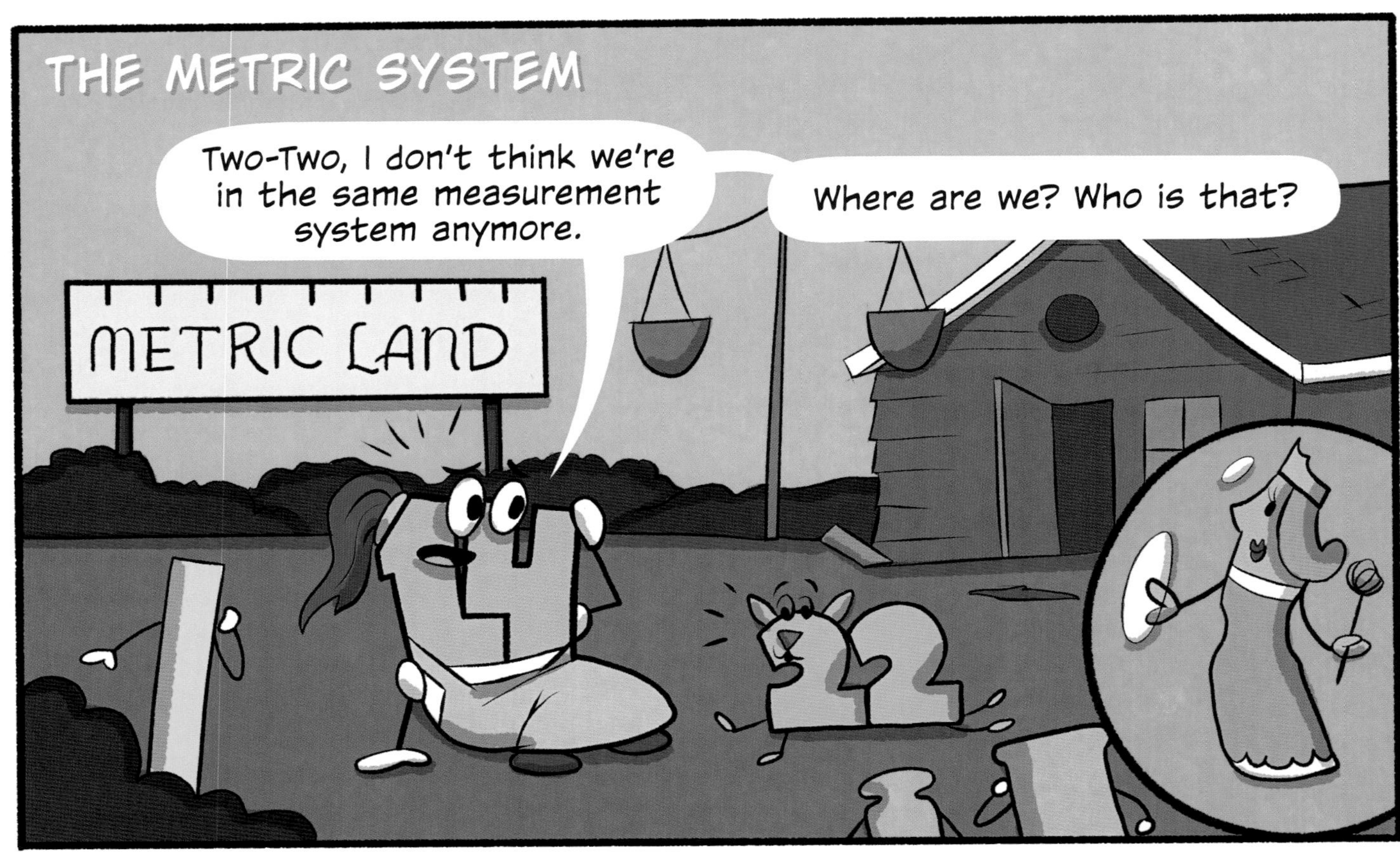
THE METRIC SYSTEM
Two-Two, I don't think we're in the same measurement system anymore.
Where are we? Who is that?
METRIC LAND

Welcome to Metric Land! I'm the Bubble Queen, your guide. The **metric system** is used all over the world to measure lengths and distances, area, volume, mass, and temperature. But in a few countries, they use the U.S. customary system.
You'll learn more about both systems soon enough.
The **gram (g)** is the basic metric unit for mass.
5g
1L
The **meter (m)** is the basic metric unit for length or distance.
The **liter (L)** is the basic metric unit for volume, especially for liquids.

The **square meter (m²)** is the basic metric unit for area.
The **cubic meter (m³)** is a unit for measuring volume.
I measure temperature in the basic metric unit of **degrees Celsius.**

I'm Dorothy. Who are you?
I am a cube with a volume of 1 cubic centimeter, or 1 cm³. Like the liter, I also measure volume, or the space that matter fills. Let me show you.

WHISTLE
!
Volume of Two-Two: 964 cm³

The Metric System			
Prefix	**Meaning**	**Example in Length and Distance Units**	**Conversion**
milli- (m)	$\frac{1}{1,000}$	1 millimeter (mm), about the width of the period at the end of a sentence	
centi- (c)	$\frac{1}{100}$	1 centimeter (cm), length of a paper clip	1 cm = 10 mm
deci- (d)	$\frac{1}{10}$	1 decimeter (dm), length of a playing card	1 dm = 10 cm
no prefix	1	1 meter (m), width of a doorway	1 m = 10 dm, or 100 cm
kilo- (k)	1,000	1 kilometer (km), about how far you can walk in 10 minutes	1 km = 1,000 m

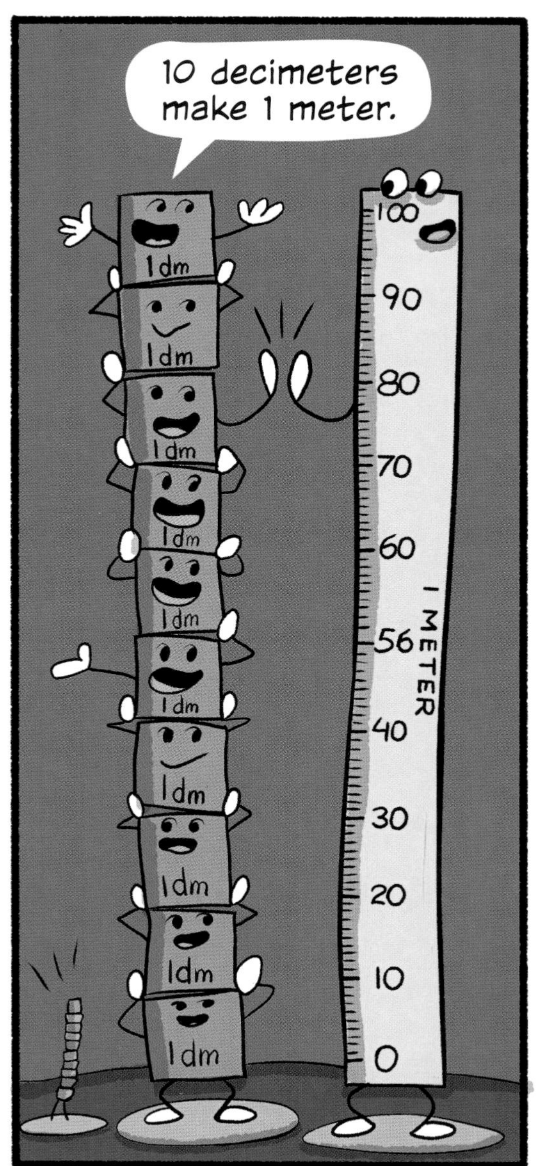

Multiply by a power of ten to convert from a large unit to a small unit. Or divide to convert from a small unit to a large unit.

The first two problems are solved for you. Can you solve the others?

A. 15 mm = ? cm — 15 ÷ 10 = 1.5, so 15 mm = 1.5 cm
B. 29 m = ? cm — 29 x 100 = 2,900, so 29 m = 2,900 cm
C. 380 km = ? m
D. 42 dm = ? m
E. 18 cm = ? mm

See page 40 for the answers.

TOOLS FOR MEASURING LENGTH OR DISTANCE

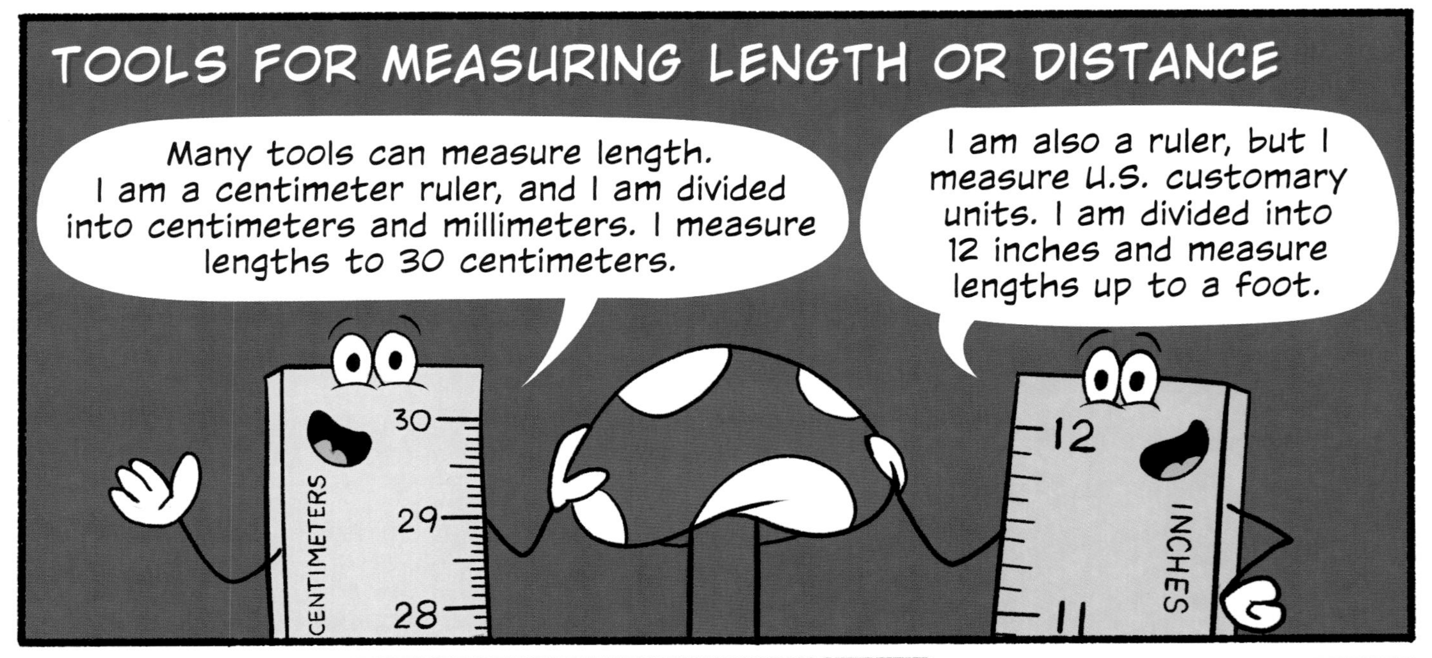

I have a device called an odometer, which shows the distance I travel in kilometers. But in countries that use the U.S. customary system, odometers show distances in miles.
...13

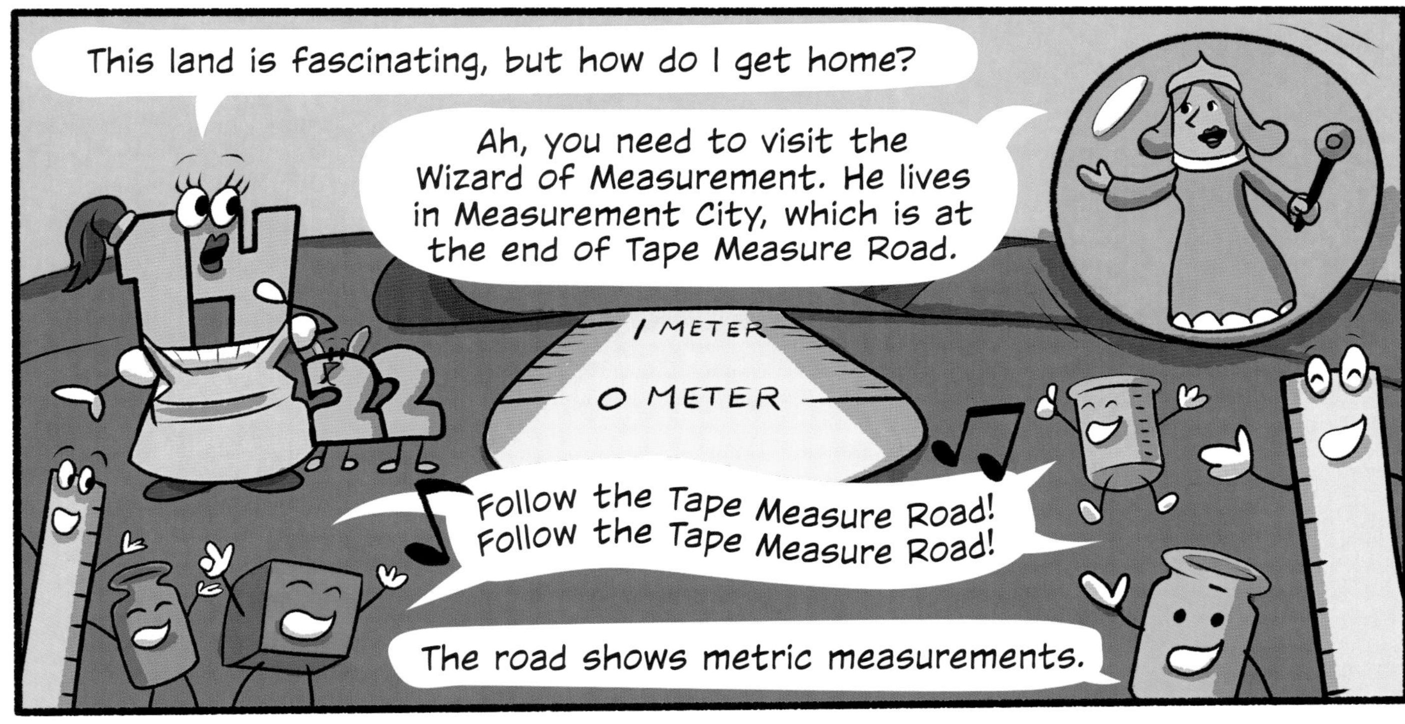
This land is fascinating, but how do I get home?
Ah, you need to visit the Wizard of Measurement. He lives in Measurement City, which is at the end of Tape Measure Road.
1 METER
0 METER
Follow the Tape Measure Road! Follow the Tape Measure Road!
The road shows metric measurements.

Hah! Measuring is much too difficult for a number like you—or your little dog, too!

Oh, for goodness sake, just ignore the Broken Ruler. Off you go!

MEASURING LENGTH AND DISTANCE
Metric Land: 3 kilometers (1.9 miles)
Measurement City: 80 kilometers (50 miles)
Height: 4 meters (13 feet)
Hold still, will you?!
What's wrong?
Hello! I'm Scarecrow, and my job is to measure the corn and the crows.
But everyone here uses metric measurements, so I'm having some trouble.
I remember the metric units for length and distance that Bubble Queen showed us!
The units are millimeters, centimeters, meters, and kilometers.
But ... my ruler uses the **United States customary system.**
That's the system my family uses! I'm Dorothy, and I'm from the United States.

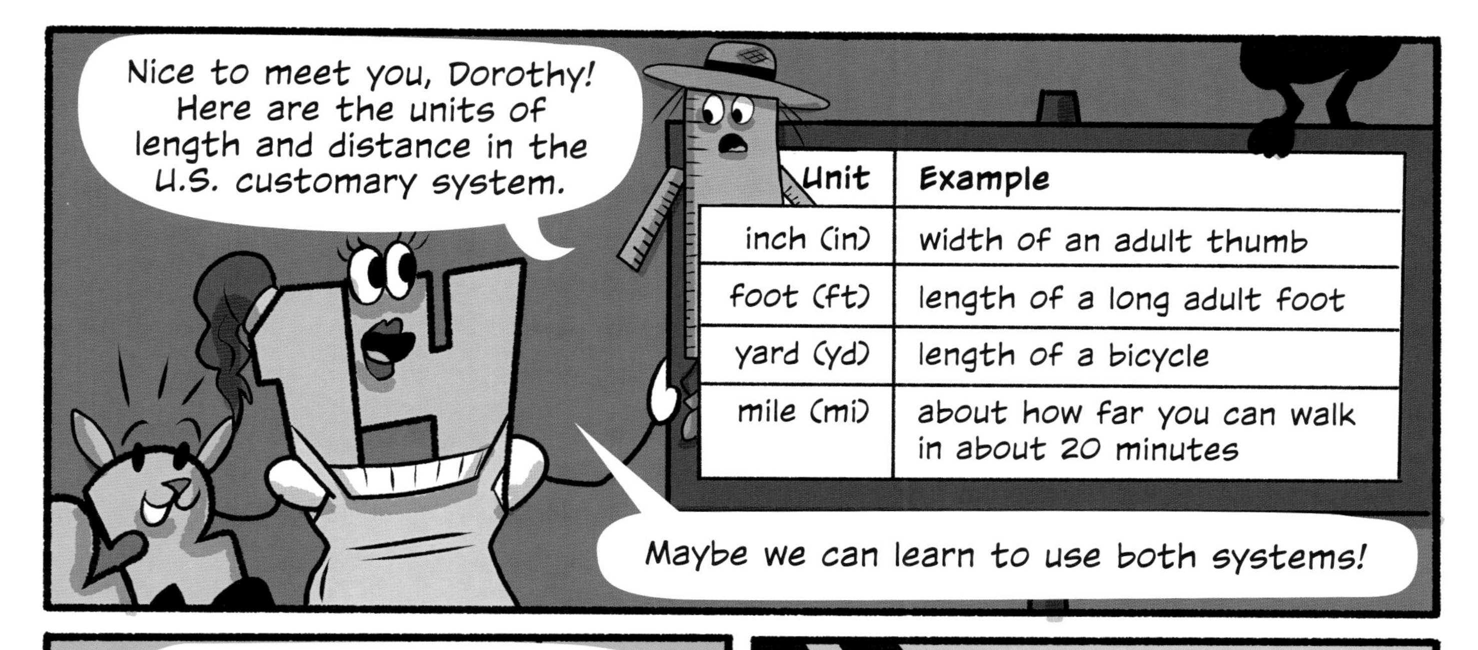

Unit	Example
inch (in)	width of an adult thumb
foot (ft)	length of a long adult foot
yard (yd)	length of a bicycle
mile (mi)	about how far you can walk in about 20 minutes

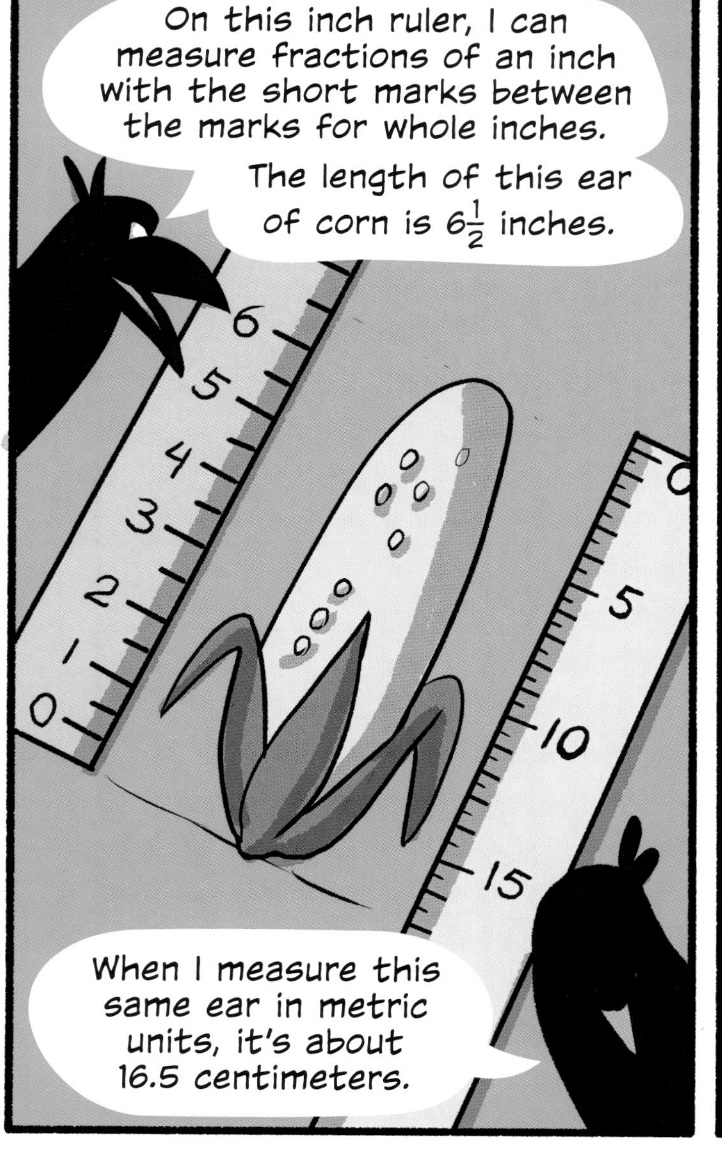

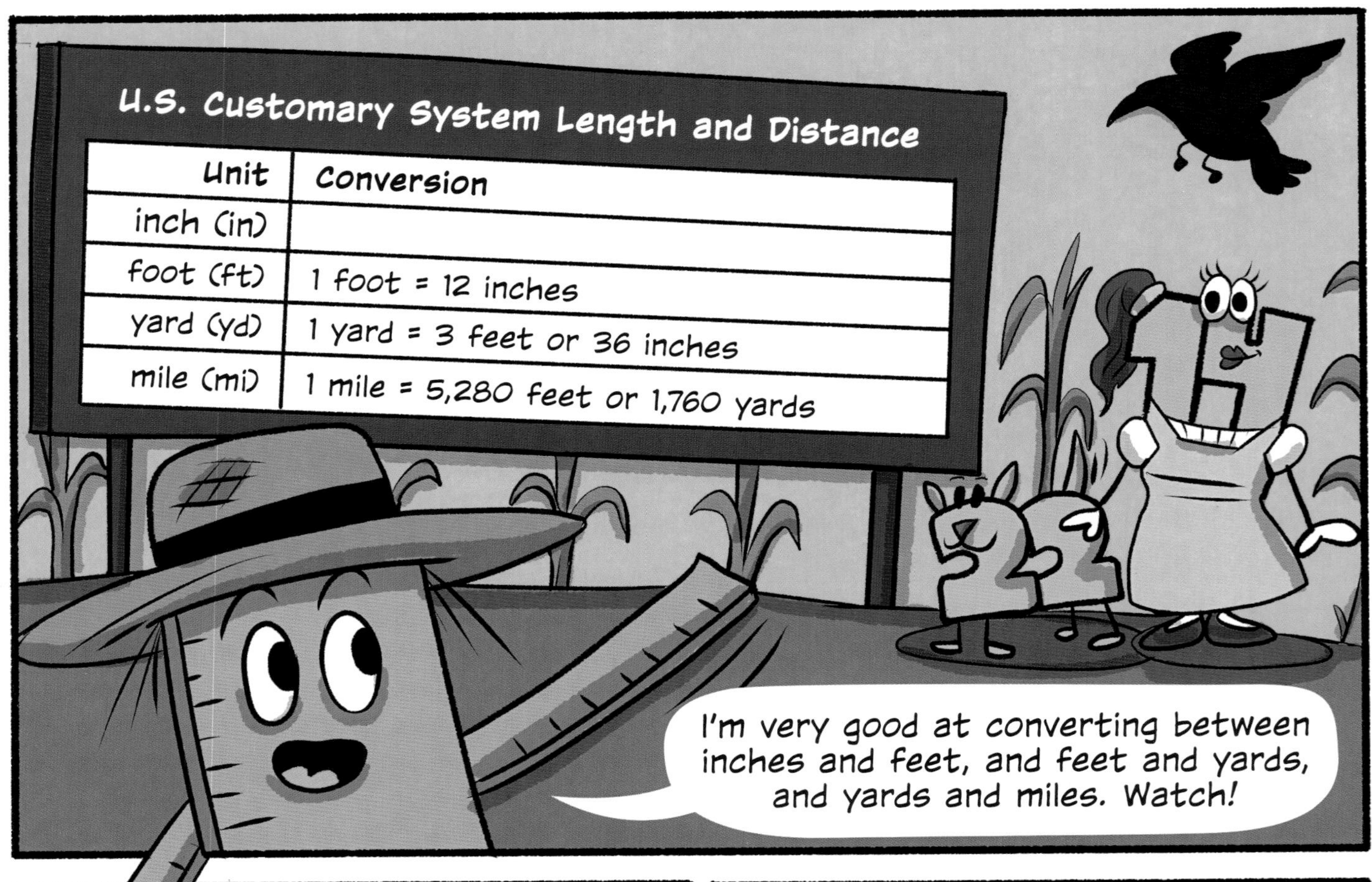
U.S. Customary System Length and Distance
Unit | Conversion
inch (in)
foot (ft) | 1 foot = 12 inches
yard (yd) | 1 yard = 3 feet or 36 inches
mile (mi) | 1 mile = 5,280 feet or 1,760 yards
I'm very good at converting between inches and feet, and feet and yards, and yards and miles. Watch!

When I find myself in pinches, With feet to change to inches, I multiply by 12!
HEIGHT = 4 FEET
×12=
48 INCHES
WINGSPAN: 24 INCHES
÷12 =
2 FEET
Or to change inches to feet, I divide by 12, it's really neat!

SKATING RINK
80 YARDS AHEAD
X3=240 FEET
To change from yards to feet.
The step just can't be beat.
I multiply by 3, you see.
And miles to feet, that's really weighty ... I multiply by 5,280!
MEASUREMENT
CITY:
50 MILES
So 50 miles is
264,000 feet,
Dad!
Well done, Three!
You are delightful, but now I must
carry on to Measurement City, home
of the Wizard of Measurement.
I'm hoping he
will teach me about
measurement and help
me get back home.
Hey! May I come with you? Maybe the Wizard
can grant me my wish of becoming a wiz at
measuring length in the metric system!

MEASURING AREA

1 METER
1 SQUARE METER

CLICK

AREA = 24 SQUARE METERS
You can measure area by the number of unit squares, like me, that cover a flat space.
1 SQUARE METER
1 METER
You can measure or use formulas to find the area of any flat surface—even circles, triangles, or irregular shapes. But the measurement is always in square units.

Welcome to the Town of the Square People. Since we are squares, we are fond of measuring area in square units. We teach all our visitors about this!
3 m
AREA = 6 m²
2 m
You can measure area in both metric and U.S. customary units. Notice that a unit for area is the square of a unit for length.
You can abbreviate the square unit with an exponent of 2, such as square centimeters (cm²), square inches (in²), and all other square units.
32 km
Area of stamp:
6 square centimeters (cm²)
or 0.9 square inch (in²)
Area of postcard:
150 square centimeters (cm²)
or 24 square inches (in²)
Dear Aunt M and Uncle H,
I am on the Tape Measure Road learning all about measurement, including area. See you soon–I hope!
Love, Dorothy
FANTASY TO REALITY MAIL

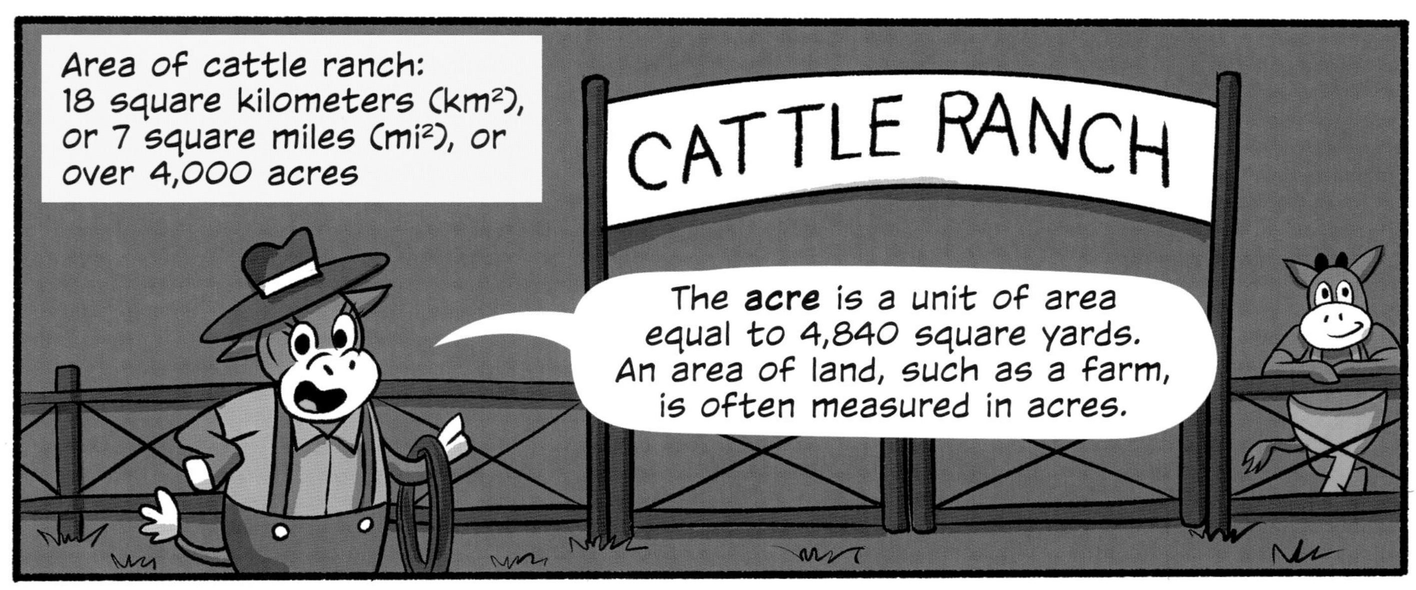
Area of cattle ranch:
18 square kilometers (km^2), or 7 square miles (mi^2), or over 4,000 acres
CATTLE RANCH
The **acre** is a unit of area equal to 4,840 square yards. An area of land, such as a farm, is often measured in acres.

Area of mattress
2.5 square meters (m^2) or 27 square feet (ft^2)
MATTRESS SALE
OPEN

PIZZA PARLOR
Area = 3.6 m^2 or 39 ft^2

Area = 45 m^2 or 54 yd^2
Give up, Dorothy
BAHAHA!
HAHAHA!

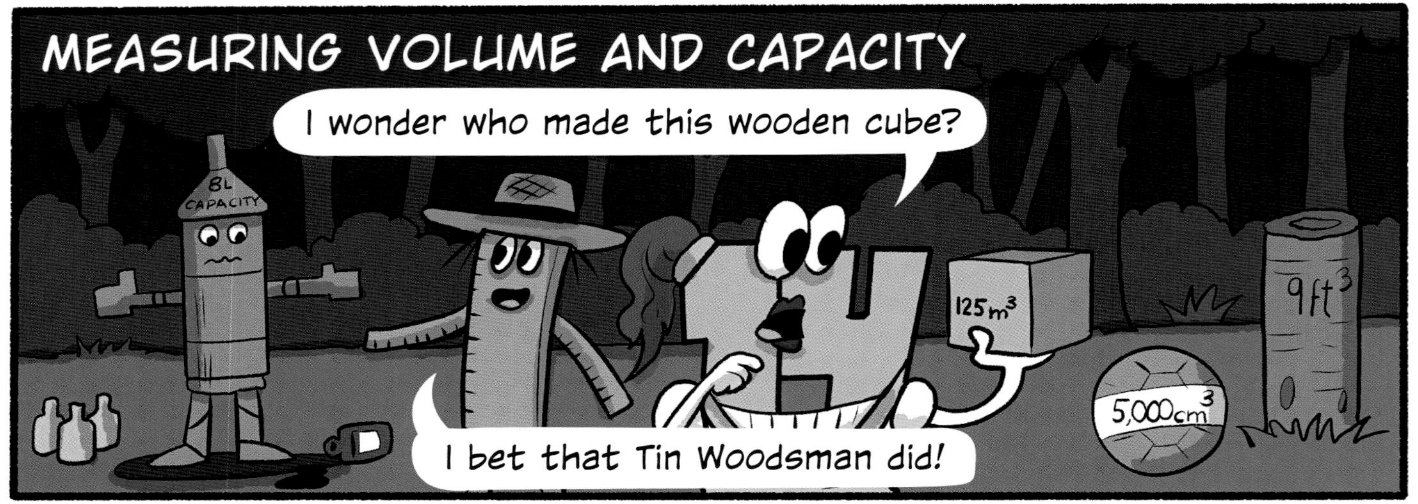
MEASURING VOLUME AND CAPACITY
I wonder who made this wooden cube?
8L CAPACITY
125 m³
9 ft³
5,000 cm³
I bet that Tin Woodsman did!

It appears that this guy needs some oil!
GLUG! GLUG!
8L CAPACITY
OIL

Oh, thank you so much. I'm a woodworker. I carve those cubes for fun.
Then why do you look rather upset?
8L CAPACITY
OIL

That's because I filled up yesterday with a volume of 8 liters of oil, but it all leaked out! Thank goodness you came along!
Volume? Liters? What are they?
8L CAPACITY

8L CAPACITY
If you've got space to measure,
For business or for pleasure,
Try quarts, gallons, liters,
Or cubic centimeters:
They're the units of **volume**!

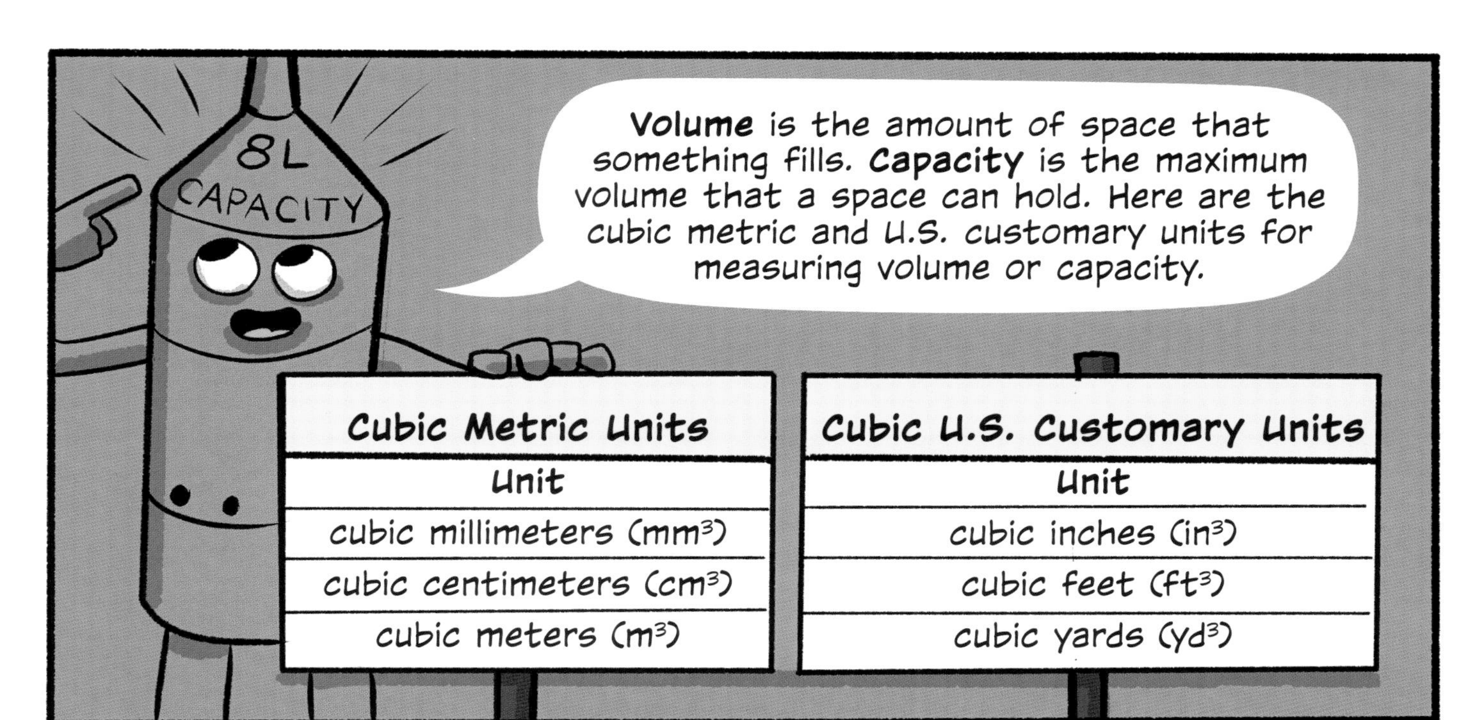

Cubic Metric Units
Unit
cubic millimeters (mm^3)
cubic centimeters (cm^3)
cubic meters (m^3)

Cubic U.S. Customary Units
Unit
cubic inches (in^3)
cubic feet (ft^3)
cubic yards (yd^3)

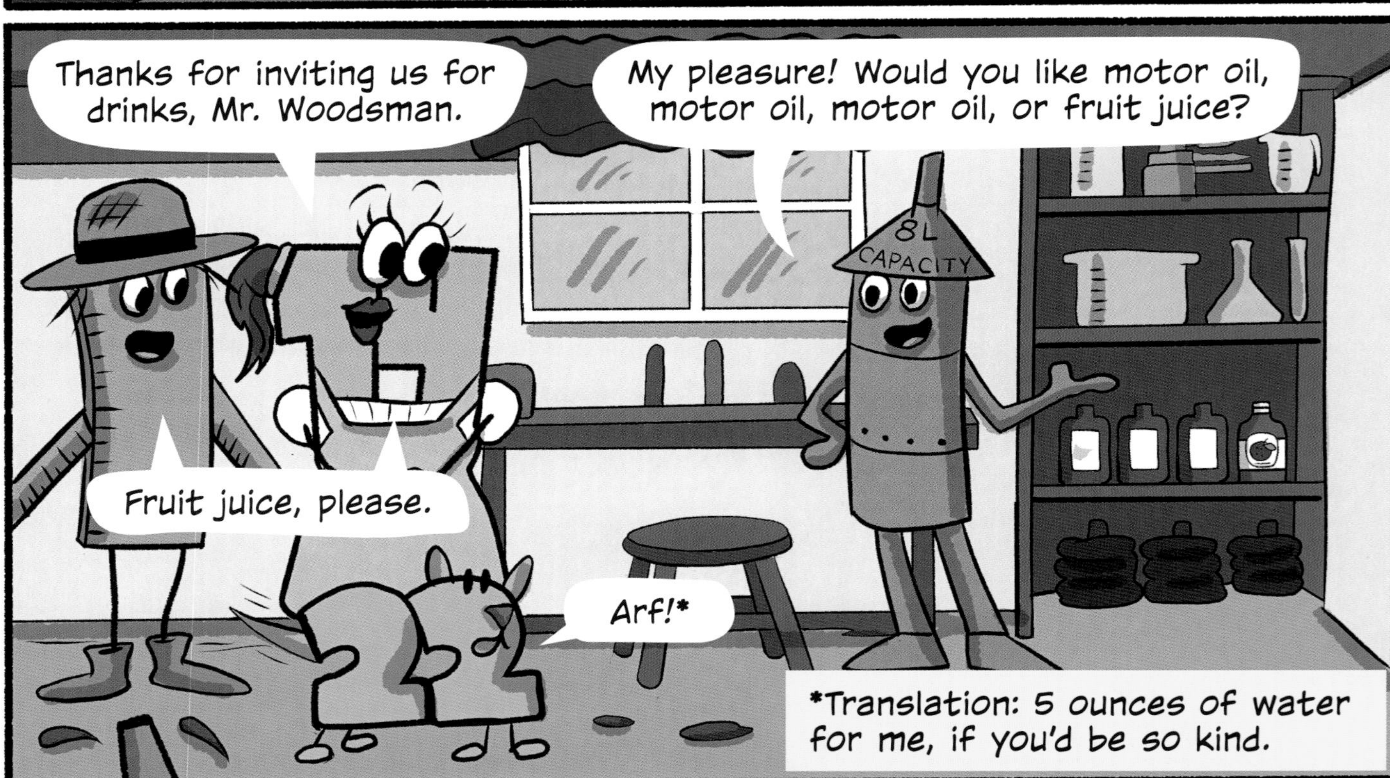

8L CAPACITY

I told you about cubic units useful for measuring volume. Here are other units for measuring volume or capacity.

U.S. Customary Units	
Unit	**Conversion**
ounce (oz)	-
cup (C)	1 C = 8 oz
pint (pt)	1 pt = 2 C
quart (qt)	1 qt = 2 pt
gallon (gal)	1 gal = 4 qt

Metric Units	
Unit	**Conversion**
milliliter (mL)	-
liter (L)	1 L = 1,000 mL

GLUG!
GLUG!
8L CAPACITY
MOTOR OIL
A liter in the metric system is about the same as a quart in the U.S. customary system.
SLURP!
I have 250 milliliters of straw-flavored juice. In liters, that's the same as 250 ÷ 1,000, or 0.250 liters. Using a U.S. customary unit of measurement, that's about 1 cup!
1,000 milliliters = 1 liter

This jug has $\frac{1}{2}$ gallon of orange juice. In quarts, that's the same as $\frac{1}{2}$ x 4 quarts, or 2 quarts.
1 gallon = 4 quarts
LAP! LAP!
1 ounce of water is about 30 milliliters.

We're headed to Measurement City to meet the Wizard there. We're hoping he grants our measurement wishes. Want to come?
Sure! I wish to know more about measuring volume so I can keep myself filled with oil.
8L CAPACITY
Dad, who drinks straw-flavored juice?
Well, Three, it's got to be tastier than motor oil.

MEASURING MASS AND WEIGHT

Mass and **weight** measure slightly different properties. But in daily life, both show how heavy or light something is. The **gram** is the basic metric unit of mass. The **pound** is the basic U.S. customary unit of weight.

Mass (Metric units)		
Unit	**Example of This Mass**	**Conversion**
1 milligram (mg)	a small feather	-
1 gram (g)	a stick of gum	1 g = 1,000 mg
1 kilogram (kg)	a small watermelon	1 kg = 1,000 g
1 metric ton (mt)	a great white shark	1 mt = 1,000 kg

Weight (U.S. Customary units)		
Unit	**Example of This Weight**	**Conversion**
1 ounce (oz)	a pencil	-
1 pound (lb)	a can of soup	1 lb = 16 oz
1 ton (t)	a small car	1 t = 2,000 lb

Average monkey:
9 kilograms or
about 20 pounds

Large raindrop:
30 milligrams or
0.001 ounce

3 female grizzly bears:
1 metric ton or 2,204 pounds

Common frog:
30 grams or
about 1 ounce

Tree:
9,000 kilograms or
10 tons (U.S. customary)

MEASURING TEMPERATURE
Aren't you forgetting an important measurement, my dear?! Bwaa ha ha ha!
Behold, it's now cold and snowy.
Or would you prefer hot and dry?!
A thermometer can measure temperature in degrees **Celsius**, for the metric system, and in degrees **Fahrenheit**, for the U.S. customary system.
°C
°F
100 °C
212 °F
water boils
38 °C
100 °F
really hot weather
21 °C
70 °F
pleasant weather
0 °C
32 °F
water freezes
-12 °C
10 °F
really cold weather

Can you change the temperature for us?
KNOC
KNOC
What would you prefer?
Refrigerator: 4 °C (40 °F)
Baking bread: 200 °C (390 °F)
The sun: 5500 °C (9900 °F)
I think we'll just walk on to the next scene. Lion, do you want to join us?
Did I hear that you're going to Measurement City and that the Wizard might be granting wishes?

That's right.
Count me in! I've got a wish, too.
8L
CAPACITY

MEASURING CIRCLES AND ANGLES
To enter, you must show that you understand measurements of the angles in these circles. Press a 72° wedge.
At last, we made it to Measurement City!
But how do we get inside?
CREAK
360°
8L CAPACITY

The first thing I notice is the label of 360° at the center of the circle. A 360° angle makes a complete circle around a point.
Knowing that, we can find measurements of other angles with vertices at the center of the circle (**central angles**) by dividing 360° into equal parts.
An **angle** is a geometric figure formed by two rays that meet at a common endpoint.
360°

This circle is divided into halves, so 360° divided into 2 equal sections is an angle of 180°.
360° divided into 3 equal sections is an angle of 120°.
8L CAPACITY
180°
CREAK
120°
Divide 360° by 4 to get 90°.
360°
72°
90°
And divide 360° by 5 to get...72°!
PRESS
72°
72°
Ah, you figured out the measurements of the central angles in all five circles. Congratulations! I'm the Gatekeeper, and I welcome you to Measurement City!
136°
99°
125°
And Dad, I remember that the sum of the central angles will always equal 360°.
Three, circles can be partitioned into angles of any sizes, not just equal sizes.

MEASURING AND COMPARING ANGLES
HALL OF ANGLES
Circles and angles are very important here in Measurement City. We use a tool called a **protractor** to measure angles precisely.
My angle is a little smaller than the **right angle**, so it measures a little less than 90°.
Your green-shaded angle measures 60°.
8L CAPACITY

An **acute angle** has a measure less than 90°.
An **obtuse angle** has a measure between 90° and 180°.
40°
140°
140°
40°
A **straight angle**, or straight line, has a measure of 180°. My lower angle is almost a straight line, so its measure is a little less than 180°.
A **reflex angle** has a measure of greater than 180°.
An **arc** is a section of the **circumference** (curved edge) of a circle. Every arc can be defined by a central angle. This is a 135° arc.
THE WIZARD
This is a 75° arc.

SOLVING MEASUREMENT PROBLEMS

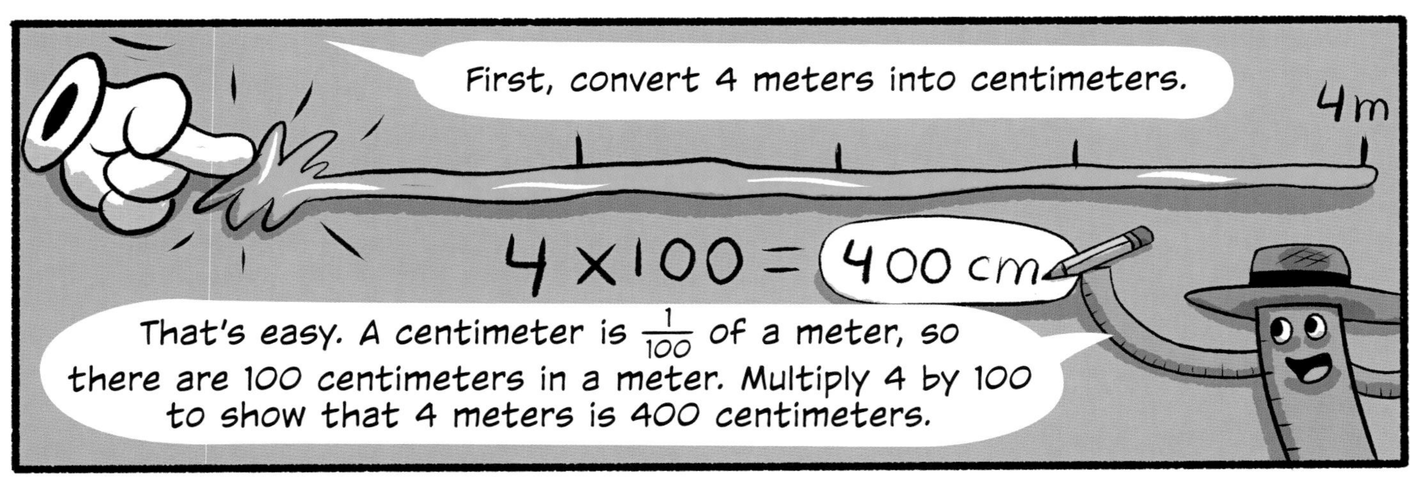

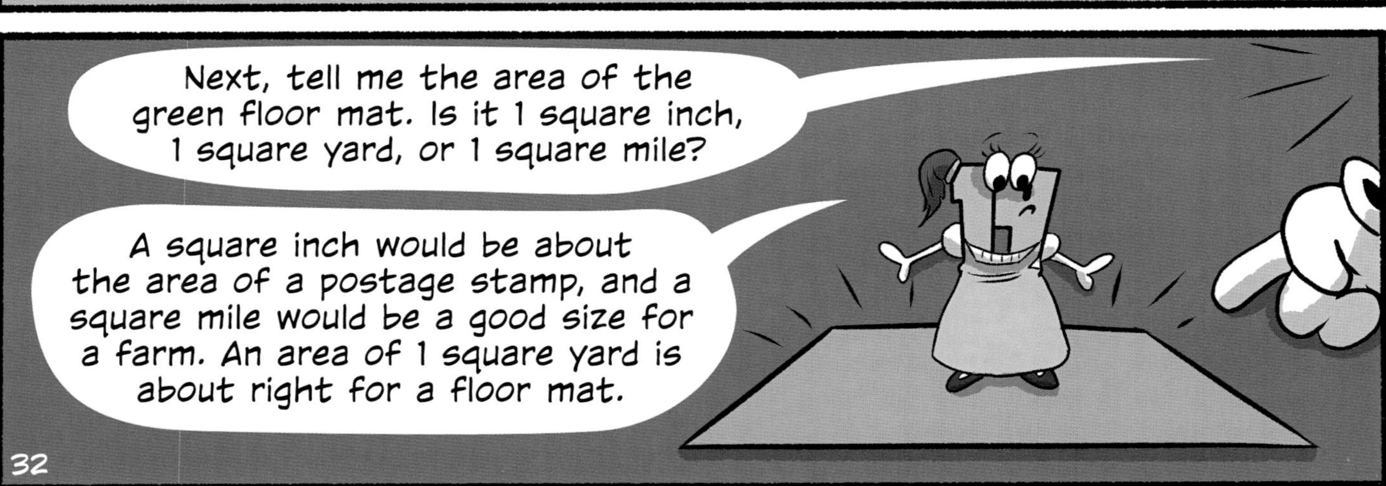

Third question,
?
!

how can 5 liters, 5 kilograms, and 50 centimeters all be measurements for this bucket of water?

The liter is a measure of volume. So, the water has a volume of 5 liters.
The kilogram measures mass, so the mass of the water is 5 kilograms.
8L CAPACITY
A centimeter measures length, so the height of the bucket could be 50 centimeters.

Good job, team.
SPLASH!
Yes, you have learned a great deal about measuring.
Aaaaahhh!!!

PUTTING MEASUREMENT TO GOOD USES
For your next tasks, you must measure the heights and weights of every tree in the forest,
and the volume of water in the ocean, and...
TUC!

Wiz, you are so busted!
TAP TAP

Yeah, now help me measure length in the metric system!
8L CAPACITY
I want to measure mass and weight without being so scared!
And help me measure volume!

Your wish is granted, my bucolic friend. We just mark millimeters and centimeters on the other side of your ruler. Now your ruler measures length in both systems.

To grant your wish, I will plug up your leaks. Now you can measure volume precisely. You will always hold 8 liters.
8L CAPACITY
ZZT!
1 LITER OF OIL
Just use a container marked with liters, and you'll have the right amount of oil on hand.

You can use an electronic scale to measure mass or weight. No fuss, no wobbling up and down on a balance scale, and no dizziness, but-
140 GRAMS
batteries not included. Your wish is granted!

Are you going to teach me everything I need to know about measurement?
You already have learned it, my dear. And if you want, you can review any of the lessons.

That goes for you, too!

TIME TO GO HOME
Now how am I going to get home?
I can take you in my hot-air balloon.
Or, just click your heels together three times.

I am a professional engineer. I can launch you with a giant rubber band,
and yes, Two-Two, too!
Let's go with the giant rubber band. That sounds pretty cool.

Distance to Kansas: 23,348 kilometers (14,508 miles);
Length of rubber band: 18.2 meters (59.7 feet);
Angle at vertex of the rubber band: 108°.
Wow, measurements certainly are important in science, technology, and engineering.
Arf, arf!*
And math, too.
*Translation: Not to mention air travel!

TWANG

Hmm, 98.6 °F. Yup, you're back to normal!
THE END

Dad, that was a great story. But let me ask one question.
Sure, son, what is it?

May I keep Two-Two?

SHOW WHAT YOU KNOW

1. Convert the following lengths.

A. 7 feet = _____ inches

B. 15 feet = _____ yards

C. 3 miles = _____ feet

D. 108 inches = _____ yards

2. Arf!*

*Translation: Convert the following volumes.

A. 4 cups = _____ ounces

B. 4,500 milliliters = _____ liters

C. 20 quarts = _____ gallons

D. 7 liters = _____ milliliters

3. Match the prefixes with their meaning.

1. centi-	A. 1,000
2. deci-	B. $\frac{1}{1,000}$
3. kilo-	C. $\frac{1}{100}$
4. milli-	D. $\frac{1}{10}$

4. Multiple choice.
Choose the best answer for each question.

1. Which of these is a unit of area?

   A. acre
   B. Fahrenheit
   C. ounce

2. Which of these is NOT a unit of volume?

   A. pint
   B. quart
   C. ton

3. Which of these is a metric unit of mass?

   A. gram
   B. pound
   C. ton

4. What does the sum of the central angles of a circle always equal?

   A. 180°
   B. 360°
   C. 720°

See page 40 for answers.

ANSWERS

Page 9:

A. 15 mm = 1.5 cm
B. 29 m = 2,900 cm
C. 380 km = 380,000 m
D. 42 dm = 4.2 m
E. 18 cm = 180 mm

SHOW WHAT YOU KNOW
Pages 38-39:

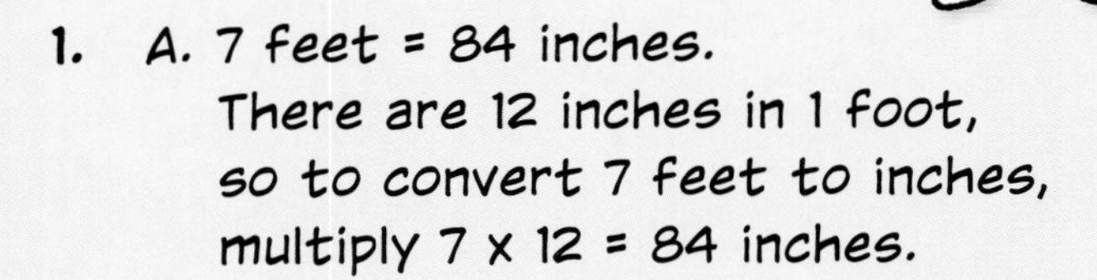

1. A. 7 feet = 84 inches.
There are 12 inches in 1 foot, so to convert 7 feet to inches, multiply 7 x 12 = 84 inches.

B. 15 feet = 5 yards
There are 3 feet in 1 yard, so to convert 15 feet to inches, divide 15 ÷ 3 = 5 yards.

C. 3 miles = 15,840 feet
There are 5,280 feet in 1 mile, so to convert 3 miles to feet, multiply 3 x 5,280 = 15,840 feet.

D. 108 inches = 3 yards

2. A. 4 cups = 32 ounces
There are 8 ounces in 1 cup, so to convert 4 cups to ounces, multiply 4 x 8 = 32 cups.

B. 4,500 milliliters = 4.5 liters
There are 1,000 milliliters in 1 liter, so to convert 4,500 milliliters to liters, divide 4,500 ÷ 1,000 = 4.5 liters.

C. 20 quarts = 5 gallons
There are 4 quarts in 1 gallon, so to convert 20 quarts to gallons, divide 20 ÷ 4 = 5 gallons.

D. 7 liters = 7,000 milliliters
There are 1,000 milliliters in 1 liter, so to convert 7 liters to milliliters, multiply 7 x 1,000 = 7,000 liters.

3. 1. C, centi- is the prefix for $\frac{1}{100}$

2. D, deci- is the prefix for $\frac{1}{10}$

3. A, kilo- is the prefix for 1,000

4. B, milli- is the prefix for $\frac{1}{1,000}$

4. Multiple choice:

1. The acre (A) is a unit of area. Fahrenheit is a unit of temperature, and ounce is a unit of weight.

2. The ton (C) is NOT a unit of volume. Ton is a unit of weight.

3. The gram (A) is a metric unit of mass.

4. The sum of the central angles of a circle always equals 360° (B).